Weather Wonders

RAINING FISH AND FROGS

Scott Wilken

Big Buddy Books

An Imprint of Abdo Publishing
abdobooks.com

abdobooks.com

Published by Abdo Publishing, a division of ABDO, PO Box 398166, Minneapolis, Minnesota 55439.

Printed in the United States of America, North Mankato, Minnesota
052025
092025

Design: Elena Klinkner, Mighty Media, Inc.
Production: Mighty Media, Inc.
Editor: Ruthie Van Oosbree
Cover Photograph: RimenPix/Shutterstock
Interior Photographs: Anterovium/Adobe Stock, p. 28 (duct tape); Carrie A Hanrahan/Shutterstock, pp. 10–11; C. Calvin/NCAR/UCAR Image and Multimedia Gallery, p. 25; Coprid/Adobe Stock, p. 28 (washers); Elena Arkadova/Shutterstock, p. 5; Elizaveta Galitckaia/Shutterstock, pp. 16–17; ellepistock/Shutterstock, p. 14; Henri Koskinen/Shutterstock, p. 23; Housework/Adobe Stock, p. 28 (Hot glue gun); Jacintne Udvarlaki/Shutterstock, pp. 26–27; Konrad Lykosthenes/Wikimedia Commons, p. 7; kyslynskahal/Shutterstock, p. 19; Michelle Lee Arnold/Adobe Stock, p. 28 (food coloring); Mighty Media, Inc., p. 29 (project); Minerva Studio/Shutterstock, pp. 12–13; Sara Winter/Shutterstock, pp. 20–21; showcake/Adobe Stock, p. 8; Soho A studio/Adobe Stock, p. 29 (water bottles); W. de Vries/Shutterstock, pp. 6–7
Design Elements: Mighty Media, Inc.

Library of Congress Control Number: 2024948582

Publisher's Cataloging-in-Publication Data
Names: Wilken, Scott, author.
Title: Raining fish and frogs / by Scott Wilken
Description: Minneapolis, Minnesota : Abdo Publishing, 2026 | Series: Weather wonders | Includes online resources and index.
Identifiers: ISBN 9781098296391 (lib. bdg.) | ISBN 9798384917823 (ebook)
Subjects: LCSH: Rain and rainfall--Juvenile literature. | Storms--Juvenile literature. | Weather--Juvenile literature. | Sky--Juvenile literature.
Classification: DDC 551.6--dc23

Contents

Animals in the Sky 4
Sightings Around the World 6
Are the Stories True? 10
Waterspouts 12
Updrafts 16
Weird Rain 20
Extreme Hail 24
Wondrous Weather 26
Make a Waterspout in a Bottle! 28
Glossary 30
Online Resources 31
Index 32

Animals in the Sky

Have you ever heard someone say "It's raining cats and dogs?" It doesn't mean cats and dogs are really falling from the sky. But **extreme** weather conditions can cause unusual objects to fall from the sky. These may include fish, frogs, and other objects!

When someone says "It's raining cats and dogs," it means that it's raining very hard.

Sightings Around the World

There have been reports of animals raining from the sky for thousands of years. In the 100s BCE, Greek **philosopher** Heraclides Lembus wrote that it rained frogs. He said houses and roads were filled with frogs.

A 1557 drawing shows a rain of frogs.

In 2007, people in Jennings, Louisiana, reported that clumps of earthworms rained from the sky.

People in Yoro, Honduras, say that it has rained fish there at least once a year since the 1800s. They have a yearly **festival** to celebrate it. In 2005, people in northwestern Serbia reported that thousands of tiny frogs rained down from the sky.

Are the Stories True?

There has not been a scientifically proven **instance** of raining fish, frogs, or other animals. People may see fish or frogs on the ground after a storm. But the animals could have been washed out of nearby bodies of water. However, some scientists believe they could have fallen from the sky.

Heavy rains and flooding may wash animals out of lakes and rivers.

Waterspouts

One **theory** is that waterspouts carry animals from one location to another. Waterspouts are cones of **rapidly** spinning air and water. Some waterspouts are **tornadoes** that occur over bodies of water.

Waterspouts may form under storm clouds.

It has never been proven that waterspouts can cause a rain of animals.

A waterspout may be able to pick up water and small objects in it. Small animals could get sucked up by strong waterspouts. Some waterspouts move from water to land. The animals would fall to the ground once the waterspouts weakened over land.

Updrafts

Another **theory** is that animals could be carried by strong updrafts. An updraft is a **current** of air that rises up from near the ground. It can reach speeds of more than 100 miles per hour (161 kmh)!

Updrafts can be caused by thunderstorms, temperature differences, wildfires, and more.

In June 1882, frogs frozen in ice were found scattered on the ground in Dubuque, Iowa. It's possible that an updraft lifted them high into Earth's **atmosphere**, where it is very cold. The frogs may have frozen into **hail** and then fallen back to the ground!

Some frogs are light enough that a strong updraft could lift them.

Weird Rain

Rain can be unusual in other ways. In 2015, rain that looked like milk fell in Washington, Oregon, and Idaho. Particles of **sodium** had mixed with the rainwater as it fell, turning it white.

The sodium causing the milky rain was lifted from a lakebed in a dust storm.

Since ancient times there have been reports of other colors of rain, most often red, yellow, and black. In 2001, red rain fell near the coast of India. It was likely because strong winds blew **algae** into the air. The algae turned the rainwater red.

Scientists believe the algae *Trentepohlia* caused the red rain in India.

Extreme Hail

Sometimes **hail** can be **extremely** large. In July 2010, huge hailstones fell in Vivian, South Dakota. One weighed nearly 2 pounds (0.9 kg)! This was the heaviest hailstone ever recorded in the United States.

A scientist at the National Center for Atmospheric Research in Boulder, Colorado, prepares to make a model of the record-breaking hailstone.

Wondrous Weather

Rain itself is not unusual. In fact, it is always raining somewhere in the world. But throughout history there have been stories of raining animals, strangely colored rain, huge **hailstones**, and more. What is the weirdest weather you've ever seen?

In 2015, it "rained" money spiders in Australia when many money spiders traveled through the sky. They used their webs to sail up in the air.

Make a Waterspout in a Bottle!

What You Need

- 2 identical bottles
- water
- food coloring (optional)
- washer at least as wide as the opening of the bottles
- hot glue gun and glue sticks
- duct tape

What You Do

1. Fill one of the bottles about three-quarters full with water.
2. Add food coloring if you want.
3. Hot glue the washer to the top of the bottle filled with water. Hot glue the empty bottle upside down on top of the washer.

4. Tape the necks of the bottles together. Use a few layers of tape to make sure no water can leak out.
5. Over a sink, turn the bottles over and give them a quick swirl to start the water moving in a circle.
6. Watch a waterspout form as the water flows to the other bottle.

Glossary

algae (AL-jee)—plants or tiny plantlike organisms that live mainly in water.

atmosphere (AT-muhs-feer)—the layer of gases that surrounds a space object.

current—wind moving steadily in a particular direction.

extreme—far beyond the usual.

festival—a party with a special focus.

hail—pieces of ice that fall during thunderstorms.

instance—a single occurrence.

philosopher (fuh-LAH-suh-fuhr)—a person who studies ideas about life.

rapid—very fast.

sodium—a soft, waxy, silver-white scientific element. It is found in things such as salt and baking soda.

theory—an idea about how or why something happens.

tornado—strong, rotating winds that often make a funnel shape. A tornado moves over land in a narrow path.

Online Resources

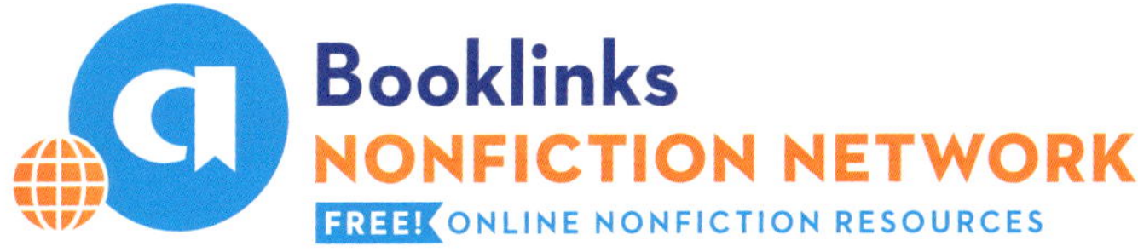

To learn more about weird rain, please visit **abdobooklinks.com** or scan this QR code. These links are routinely monitored and updated to provide the most current information available.

Index

algae, 22, 23
atmosphere, 18, 25

Colorado, 25
colors, 20, 22, 23, 26

Dubuque, Iowa, 18

fish, 4, 9, 10
frogs, 4, 6, 7, 9, 10, 18, 19

hail, 18, 24, 25, 26

Idaho, 20

Lembus, Heraclides, 6

Make a Waterspout in a Bottle! project, 28, 29
milk, 20, 21

Oregon, 20

scientists, 10, 23
Serbia, 9
sodium, 20, 21

theories, 12, 16

Trentepohlia, 23

updrafts, 16, 17, 18, 19

Vivian, South Dakota, 24

Washington, 20
waterspouts, 12, 13, 14, 15, 28, 29
wind, 16, 17, 18, 19, 22

Yoro, Honduras, 9